# {锦绣 山河}
## WONDERFUL LAND

中国王屋山—黛眉山世界地质公园
Wangwushan-Daimeishan UNESCO Global Geopark of China

中国王屋山—黛眉山世界地质公园管理委员会 编
The Administration of Wangwushan-Daimeishan UNESCO Global Geopark, China

黄河水利出版社
THE YELLOW RIVER WATER CONSERVANCY PRESS
·郑州·
Zhengzhou

穿越地球时空的地质走廊

# 三山一河织锦绣
## Three mountains and one river weaving the splendor

太行山在这里戛然而止
Here Taihangshan range ends abruptly

王屋山在这里横空出世
Here Wangwushan makes its sudden appearance

黛眉山在这里锦上添花
Here Daimeishan adds more flavor to the beauty

黄河在这里气象万千
Here the Yellow River presents her ever-changing magnificence

# 大山大河描绘的锦绣画卷

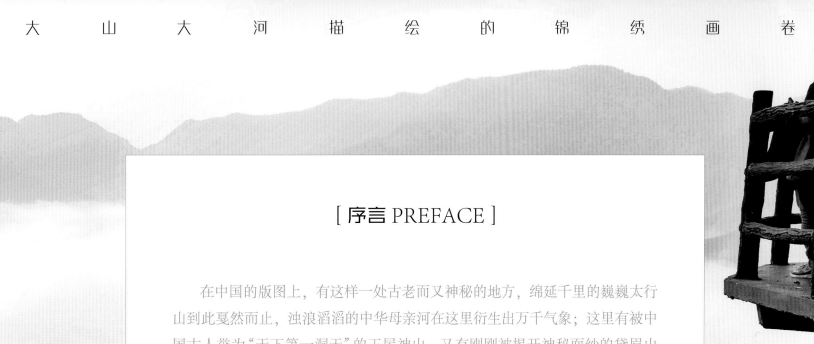

## [ 序言 PREFACE ]

　　在中国的版图上，有这样一处古老而又神秘的地方，绵延千里的巍巍太行山到此戛然而止，浊浪滔滔的中华母亲河在这里衍生出万千气象；这里有被中国古人誉为"天下第一洞天"的王屋神山，又有刚刚被揭开神秘面纱的黛眉山红岩嶂谷群；构造地质学家在这里找到了早期古大陆聚散的关键证据，地层古生物学家在这里建立了晚三叠世到早侏罗世陆相地层的典型剖面；游客在这里追寻黄河演化的历史，诗人在这里慨叹大自然的无穷魅力。

　　一条穿越地球时空的地质走廊，一幅由大山大河描绘的锦绣画卷。

　　这就是中国王屋山—黛眉山世界地质公园。

In the Chinese territory, there is a place of antiquity and mystery, where the lofty Taihangshan range ends abruptly, and the waving Yellow River-mother river of China presents her ever-changing magnificent sceneries. Here stands the sacred mountain of Wangwushan, which is eulogized by ancient Chinese as one of the most famous Taoist mountains, and the Daimeishan red stone valleys with their mysteries unveiled. Experts of tectonics found the key evidence of the assembling and breaking apart of the ancient continents in the early period on the earth. Stratigraphical experts and paleontologists established the typical section of continental facies sequence from Late Triassic to Early Jurassic epoches here. Tourists can pursue the evolution of the Yellow River while poets may be amazed at the infinitude of the nature's magic.

This is a geological corridor running through the time and space of the earth. It is also a splendor painting by the great mountains and river.

Such is Wangwushan-Daimeishan UNESCO Global Geopark, China.

United Nations  
Educational, Scientific and  
Cultural Organization

联合国教育、  
科学及文化组织

Wangwushan–Daimeishan  
UNESCO  
Global Geopark

王屋山－黛眉山  
联合国教科文组织  
世界地质公园

# 目 录 CONTENTS

I 公园概况 Introduction to the Geopark ..................... 002

II 王屋山景区 Wangwushan Scenic Area ..................... 016

III 小沟背(银河峡)景区 Xiaogoubei (Yinhe Valley) Scenic Area ..................... 032

IV 黄河三峡景区 The Yellow River Three Gorges Scenic Area ..................... 040

V 龙潭峡景区 Longtan Valley Scenic Area ..................... 052

VI 黛眉山景区 Daimeishan Scenic Area ..................... 068

VII 荆紫仙山景区 Jingzixianshan Scenic Area ..................... 082

VIII 公园人文景观 Cultural Heritage in the Geopark ..................... 090

IX 公园生态资源 Ecological Resources in the Geopark ..................... 104

X 公园旅游 Geopark Tourism ..................... 112

# I 公园概况
## Geopark Profile

　　王屋山—黛眉山世界地质公园位于中国河南省西北部的济源市与新安县境内,由王屋山、小沟背(银河峡)、封门口、黄河三峡、龙潭峡、黛眉山6个典型地质遗迹区组成,总面积986平方千米,于2006年由联合国教科文组织批准。它是一座以地质剖面、地貌景观和古生物化石地质遗迹为主,与动植物资源、历史文化相互辉映的综合型地质公园:天坛山地质构造剖面系统反映了25亿年以来的地质演化历史;小沟背(银河峡)展示了17亿年前的古火山风姿;黄河三峡的八里峡是黄河贯通形成的重要节点;黛眉山红石峡谷群红岩碧水堪称山水画廊,又是波痕、泥裂等沉积构造的天然博物馆;封门口一带的晚古生代似哺乳类动物化石和中生代的遗迹化石等具有世界对比意义。公园有植物1000余种,堪称"华北种子植物的基因库";太行猕猴憨态可掬,是山中的精灵;愚公移山精神铸就了中华民族不屈的民族魂,是宝贵的精神财富。

　　Wangwushan-Daimeishan UNESCO Global Geopark is located in Jiyuan City and Xin'an County, the north west part of Henan Province. It is composed of six typical scenic areas, such as Wangwushan, Xiaogoubei (Yinhe Valley), Fengmenkou, the Yellow River Three Gorges, Longtan Valley and Daimeishan, covering an area of 986km$^2$. The geopark was approved in 2006 by the UNESCO. It is a comprehensive geopark, featured the geoheritages as geological section, landform landscape and paleontology fossils, shining with animal and plant resources, and historical culture.

　　The system of Tiantan Mountain geological structure section reflects the geological evolutionary history since 2.5Ga, Xiaogoubei Valley (Yinhe Valley) shows the charm of 1.7Ga ago; Bali Valley on the Yellow River Three Gorges is the key point during the running-through of the Yellow River; The group of red stone valleys in Daimeishan can be said to be the gallery of mountains and waterscape, which is also the natural museum of ripple marks and mud cracks; Mammal-like fossil in Late Paleozoic and trace fossil in Mesozoic around Fengmenkou area have the significance of global correlation. More than 1000 species of plants live in the geopark, called the 'gene base of seed plants in North China'; The charmingly naive Taihang macaque is the spirit in the mountain; The spirit of Yugong moving mountains is valuable spiritual wealth, which made the unyielding national soul of China.

洞天福地
The Stele of Heaven

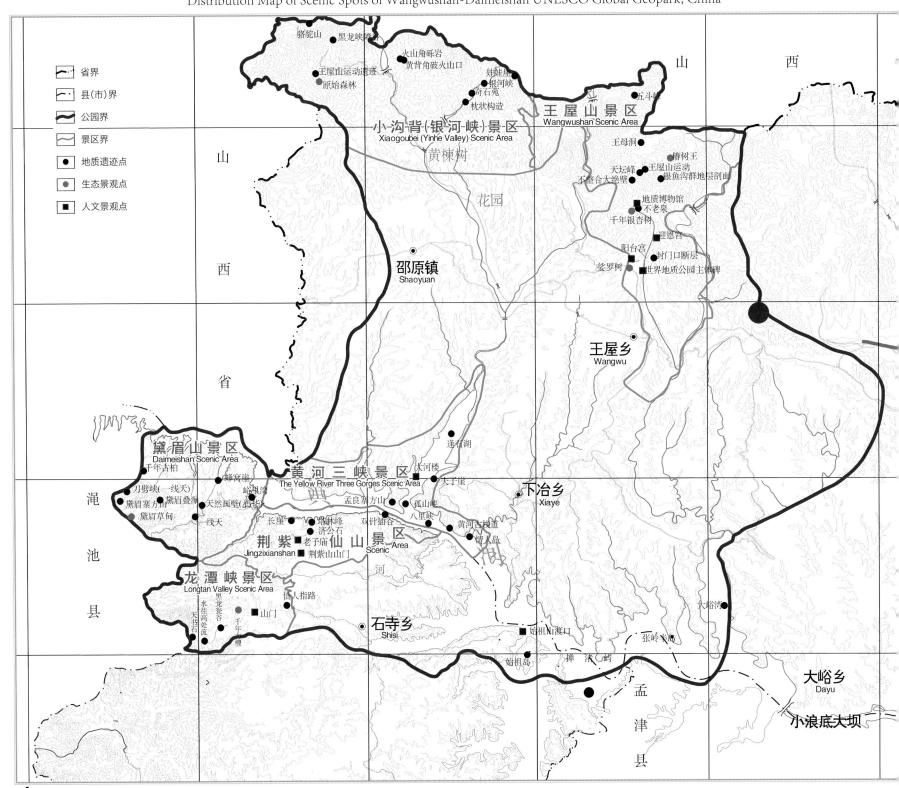

## 公园地层表
### Stratigraphical Table of Geopark

| 界 | 系 | 统 | 群 | 组 | 地层代号 | 厚度（米） |
|---|---|---|---|---|---|---|
| 新生界 Kz | 第四系 Q | 全新统 Qh | | | | 0～10 |
| | | 上更新统 $Qp_3$ | | | | 5～77 |
| | | 中更新统 $Qp_2$ | | | | 28～72 |
| | 新近系 N | 中新统 $N_1$ | | 洛阳组 | $N_1 l$ | 120 |
| | 古近系 E | 始新统 $E_2$ | 济源群 $E_2 J$ | 南姚组 | $E_2 ny$ | 190 |
| | | | | 泽峪组 | $E_2 z$ | |
| | | | | 余庄组 | $E_2 y$ | |
| | | | | 聂庄组 | $K_2 d$ | |
| 中生界 Mz | 白垩系 K | 上统 $K_2$ | | 东孟村组 | $J_3 h$ | >152 |
| | 侏罗系 J | 下统 $J_3$ | | 韩庄组 | $J_2 m$ | 21 |
| | | 中统 $J_2$ | | 马凹组 | $J_2 y$ | 250 |
| | | | | 杨树庄组 | $J_1 a$ | 120 |
| | | 下统 $J_1$ | | 鞍腰组 | $J_1 a$ | 250 |
| | 三叠系 T | 上统 $T_3$ | 延长群 $T_{2-3} Y$ | 谭庄组 | $T_3 t$ | >2000 |
| | | | | 椿树腰组 | $T_3 \hat{c}$ | |
| | | 中统 $T_2$ | | 油房庄组 | $T_2 y$ | |
| | | | | 二马营组 | $T_2 er$ | |
| | | 下统 $T_1$ | 石千峰群 $P_3-T_1 \hat{S}$ | 和尚沟组 | $T_1 h$ | 189～260 |
| | | | | 刘家沟组 | $T_1 l$ | 235～470 |
| 古生界 Pz | 二叠系 P | 上统 $P_3$ | | 孙家沟组 | $P_3 S$ | >1000 |
| | | 中统 $P_2$ | | 石盒子组 | $P_{1-2} \hat{s}$ | |
| | | 下统 $P_1$ | | 山西组 | $P_1 s$ | |
| | 石炭系 C | | | 太原组 | $C_2-P_1 t$ | >100 |
| | | 上统 $C_2$ | | 本溪组 | $C_2 b$ | |
| | 奥陶系 O | 中统 $O_2$ | | 马家沟组 | $O_2 m$ | 360 |
| | 寒武系 € | 上统 $€_2$ | | 三山子组 | $€_3 s$ | 260 |
| | | | | 炒米店组 | $€_3 \hat{c}$ | |
| | | | | 山组 | $€_3 g$ | |
| | | 中统 $€_2$ | | 张夏组 | $€_2 \hat{z}$ | 270 |
| | | | | 馒头组 | $€_{1-2} m$ | |
| | | 下统 $€_1$ | | 辛集组 | $€_2 x$ | >120 |
| 中元古界 $Pt_2$ | 蓟县系 Jx | | 汝阳群 JxR | 北大尖组 | Jxbd | 1000 |
| | | | | 白草坪组 | Jxb | |
| | | | | 云梦山组 | Jxy | |
| | | | | 小沟背组 | Jxx | |
| | 长城系 Ch | | 西阳河群 ChX | 马家沟组 | Chm | >3000 |
| | | | | 鸡蛋坪组 | Chj | |
| | | | | 许山组 | Chx | |
| | | | | 大古石组 | Chd | |
| 古元古界 $Pt_1$ | | | 银鱼沟岩群 $Pt_1 Y$ | 双房岩组 | $Pt_1 \hat{s} f$ | >3000 |
| | | | | 北崖山岩组 | $Pt_1 b$ | |
| | | | | 赤山沟沿组 | $Pt_1 \hat{c}$ | |
| | | | | 幸福园岩组 | $Pt_1 x$ | |
| 新太古界 $Ar_3$ | | | 林山岩群 $Ar_3 L$ | 迎门宫岩组 | $Ar_3 y$ | >2800 |
| | | | | 曹庄岩组 | $Ar_3 c$ | |

王屋山
Wangwushan

黛眉山高山草甸
Daimeishan Alpine Meadow

黛眉山天使之吻
Daimeishan Angel Kiss

龙潭奇观
Longtan Valley

黄河三峡风光
The Yellow River Three Gorges

# II 王屋山景区
## Wangwushan Scenic Area

王屋山景区位于王屋山—黛眉山世界地质公园北部,"愚公移山"的故事就发生在这里。王屋山是中国古代九大名山之一,唐时被列为道教十大洞天之首,称"天下第一洞天",主峰天坛山是中华民族祖先轩辕黄帝设坛祭天之所,世称"太行之脊""擎天地柱";天坛山地质构造剖面完整记录了距今25亿~14亿年前地球演化的重大地质事件,见证了当时古大陆的聚合与裂解;景区森林覆盖率在98%以上,其中最为珍奇的是有"世界植物活化石"之称、树龄达2000年以上的银杏树。

　　Wangwushan Scenic Area is located in the north part of Wangwushan-Daimeishan Global Geopark, where the story of 'Yugong moving mountains' happened. Wangwushan is one of the nine famous mountains in ancient China, which was listed as the head of the ten Taoism fairylands in Tang Dynasty. The main peak, Tiantan Peak, is the place where an altar is set up to sacrifice the Emperor Xuanyuan, the ancestor of the Chinese nation, which is called the spine of Taihang Mountains and the skyscraper by people; Tiantanshan tectonic section completely records the big geological events from 2.5Ga to 1.4Ga about the earth evolution, and witnessed the aggregation and breakup of the paleocontinent. The forest coverage rate of the scenic area reaches 98%, and the ginkgo tree, which is more than 2000 years old and called the 'living fossil among the plant world', is the most valuable variety.

王者之屋
King Palace

# 王屋山科普线路图
## Wangwushan Science Popularization Trail

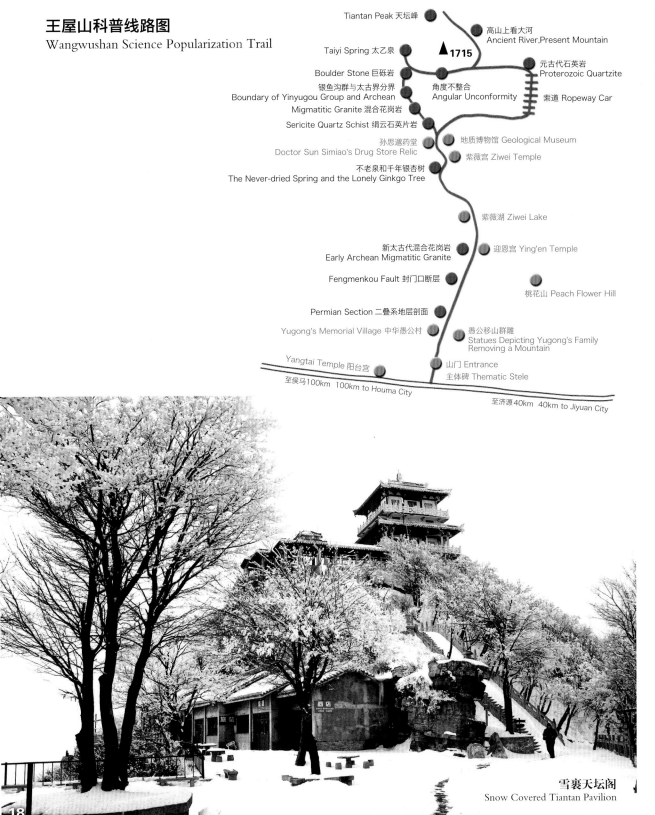

雪裹天坛阁
Snow Covered Tiantan Pavilion

迷人雾凇
Charming Rime

峰峦叠嶂王屋山
Wangwushan's Overlapping Peaks

王屋云海
Cloud Sea on Wangwushan

日映王屋
Wangwushan Reflected in the Sunshine

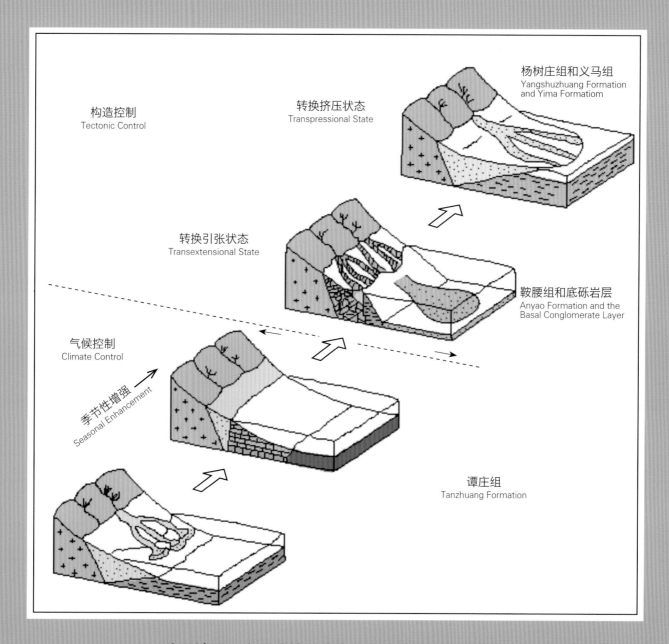

**豫西济源—义马盆地早中生代湖泊体系演化示意图**
*Evolution of the early Mesozoic Lacustrine System in Jiyuan-Yima Basin, West Henan*

天下第一洞天
Taoism Fairyland

攀岩而上
Rock-climbing

天坛山构造地质剖面实景
Tiantanshan Tectonic Section

## 天坛山构造地质剖面图
## Map of Tiantanshan Tectonic Section

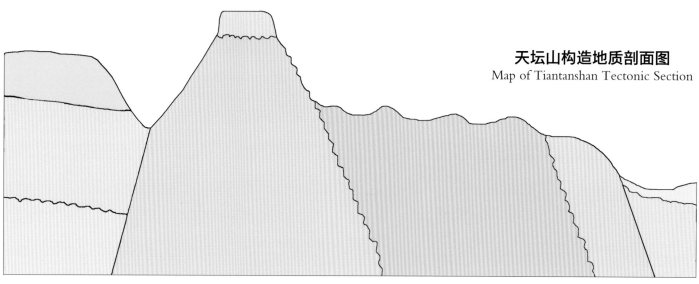

| 时代和界线(Ma)<br>Age and Boundaries | 岩石地层<br>Rock-stratigraphic Unit | 事件柱<br>Event Column | 构造岩石组合<br>Structure-Rock Assemblage | 区域构造样式<br>Regional Tectonic Pattern | 区域变质<br>Regional Metamorphism | 大地构造 发展阶段<br>Tectonic Evolution Stage | 主要测年 数据(Ma)<br>Main Dating Data |
|---|---|---|---|---|---|---|---|
| 寒武纪<br>Cambrian Period | | | 碳酸盐岩<br>Carbonates | 近水平产出<br>Nearly-horizontal | 未变质<br>Unmetamorphosed | 稳定陆块<br>Stable Continent | |
| — 800±10 — | 汝阳群<br>Ruyang Group | | 砂泥质岩<br>红色碎屑岩<br>Red Sandstones and Mudstones | | | | |
| 中元古代<br>Mesoproterozoic Era | 西阳河岩群<br>Xiyanghe Rock Suite | | 安山岩<br>Andesites | 平缓开阔褶皱<br>Broad Folds | 未变质<br>Unmetamorphosed | 三叉裂谷<br>Trident Rift | |
| — 1800±50 — | 银鱼沟岩群<br>Yinyugou Rock Suite | | 复理式碎屑岩泥质岩—碳酸盐岩建造夹中基性火山岩<br>Flysch-type Clastic Rocks-Carbonate Foranation with Intermediate to Basic Volcanic Rocks | 平卧褶皱及韧性剪切带<br>Recumbent Folds and Ductile Shear Zones | 低绿片岩相至低角闪岩相<br>Lower Greenschist to Lower Amphibolite facies | 大陆裂谷<br>Continental Rift<br>↑<br>微陆块生长<br>Micro-continent Accretion | 1840<br><br>2104±5<br>2059±5 |
| 古元古代<br>Paleoproterozoic Era | | | | | | | |
| — 2500±100 — | 林山岩群<br>Linshan Rock Suite | | 双峰式火山岩<br>Bi-modal Volcanic Rocks<br>花岗斑岩<br>Granitic Porphyries<br>复理式浊积岩系<br>Flysch Turbidites | 平卧褶皱及韧性剪切带<br>Recumbent Folds and Ductile Shear Zones | 角闪岩相<br>Amphibolite Facies | 微陆块形成<br>Micro-continent Accretion | 2744±7<br>2769±6<br>2530±3<br>2115±6 |
| 新太古代<br>Neoarchean Era | | | | | | | |

## 王屋山前寒武纪地壳演化事件柱
## Precambrian Crust Evolution Events in Wangwushan

Triassic System 三叠系

槽状交错层理
Grooved Cross-bedding

中元古界汝阳群
Mesoproterozoic Ruyang Group

鲕粒灰岩　豆粒灰岩
Oolite Limestone　Pea Limestone

**泥质条带灰岩**
Clay Streak Limestone

**皮壳状构造**
Crusty Structure

# Ⅲ 小沟背(银河峡)景区
## Xiaogoubei (Yinhe Valley) Scenic Area

　　小沟背(银河峡)景区位于公园西北部,有华北地区最典型、最壮观的火山岩峡谷地貌和最美丽的"五色石"砾岩景观。这里距今18.5亿～14.5亿年前的火山岩是世界上最古老的未变质火山岩,清晰地记录了王屋山由海变山和华北古大陆裂解、拼接的全过程,而砾岩则展示了距今14.5亿年前后本区由山间盆地发展为古代大河的全过程;火山岩形成的银河峡曲径通幽,弯弯曲曲的银河潺潺流淌,满沟的"五色石"大如楼宇,小似鸟卵,七叠瀑、斩龙潭、太极石、凤凰台、女娲庙等景观让人流连忘返。

　　Xiaogoubei (Yinhe Valley) Scenic Area is located in northwest of the Geopark, where there is the most typical and spectacular volcanic valley landform and the most beautiful 'color stone' conglomerate landform. The volcanic rocks around 1.85 billion years ago to 1.45Ga is the oldest unmetamorphosed rocks in the world, which clearly recorded the whole process from ocean to mountains of Wangwushan area and the breakup to splicing of North China paleocontinent. The conglomerate shows the process from mountain basin to paleo-river around 1.45Ga in this area; The Yinhe Valley formed by volcanic rocks flows like the curving Milky Way, and the color stones all over the valley look like big buildings or small birds. In addition, the sceneries like Sevenfold Waterfall, Zhanlong Pond, Taiji Stone, Fenghuang Platform, Nuwa Temple make people be enamoured with them.

### 小沟背(银河峡)科普线路图
### Xiaogoubei (Yinhe Valley) Science Popularization Trail

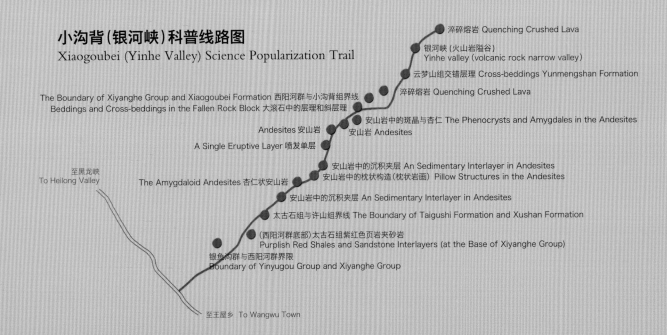

Volcanic Rock Valley

火山岩峡谷

柱状节理
columnar Joints

淬碎熔岩
Quenching Crushed Lava

火山岩峡谷
Volcanic Rock Valley

Green Mountain and Clear Water

山清水秀

枕状熔岩 Pillow Lava

古元古界银鱼沟岩群

小沟背组砾岩 Xiaogoubei Formation Conglomerate

火山岩中的沉积夹层 Sedimentary Interlayers in Volcanic Rocks

# Ⅳ 黄河三峡景区
## The Yellow River Three Gorges Scenic Area

黄河三峡景区位于黄河北岸、济源市境内。景区为发育在5亿年前寒武纪灰岩中的三条峡谷，孤山峡鬼斧神工、群峰竞秀；龙凤峡九曲十折、峡深谷幽；八里峡峭壁如削、雄伟壮观，号称"万里黄河第一峡"，是黄河中下游最后一段峡谷，对研究黄河贯通东流入海具有重要意义。

景区不仅山、水、崖、洞和谐交融，自然资源丰富，而且还有鲧山禹斧、犀牛望月、孟良活地、京娘化凤、石人顶石山、章公背章婆等自然人文景观80余处。

The Yellow River Three Gorges Scenic Area is located in Jiyuan City and the north bank of the Yellow River. The scenic area includes three gorges developed among limestone in Precambrian 500Ma ago. Gushan Gorge is the wonderfulness of the nature with peak groups contending for beauty; Longfeng Gorge is zigzagged with deep valleys and quiet mountains; Bali Gorge is steep and magnificent which is called the best gorge along the ten-thousand-mile Yellow River, which is the last valley on the middle and lower reaches of the Yellow River, having important significance on the research on the running-through and flowing east to the sea of the Yellow River. Mountains, waterscape, cliffs and caves blend harmoniously in the scenic area with rich natural resources. In addition, many natural and cultural heritages also exist in the area, namely, Mountain Gun and Yu Axe, Watching Moon Rhinoceros, Mengliang Huodi, Jingniang Huafeng, Shirending Stone Mountain, Grandfather Zhang Bearing Grandmother Zhang. The number of the scenic spots in the area reaches 80.

黄河三峡风光
The Yellow River Three Gorges Scenery

高峡平湖
High Canyons Rising out of the Lake

# 黄河三峡科普线路
## The Yellow River Three Gorges Science Popularization Trail

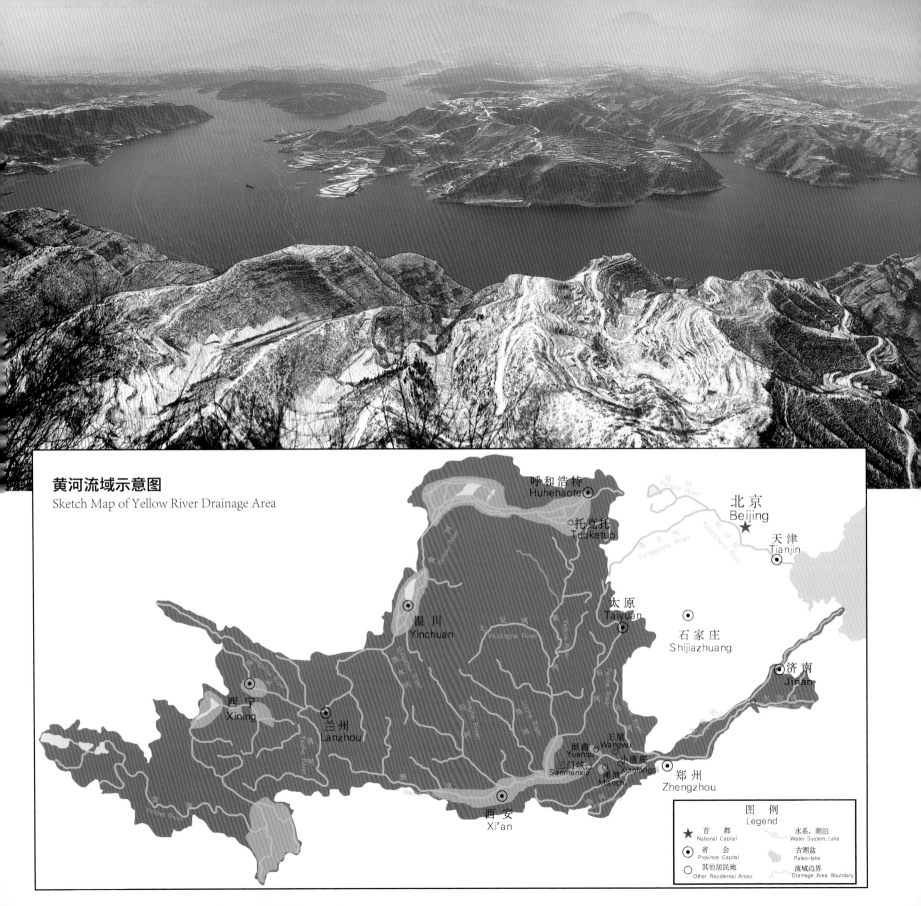

雪染黄河
Snow-colored the Yellow River

峡湾风光
Valleys Scene

## 风水宝地
Precious Place of Great Mountains and Water

## 十里画廊
Thousands of Metres Long Picture Corridor

黄河第一峡——八里峡。八里峡位于荆紫仙山脚下，黄河中下游最后一道峡谷，全长4000米，宽约200米。两岸悬崖峭壁，奇峰异石，峡内石岛林立，山环水抱，呈现出北国少见、雄浑壮丽的山光水色，有十里画廊之美誉。

Bali Gorge, situated at the foot of Jingzixianshan, is the last valley along the middle-lower reaches of the Yellow River, 4000m long and 200m wide. Here steep clipps, magnificent peaks, grotesque stones and clear water encircling small islands, present splendid and grandiose scenes which are rarely seen in North China, which win the reputation of 'Ten-mile Gallery'.

黄河冬景
Winter Scenery of the Yellow River

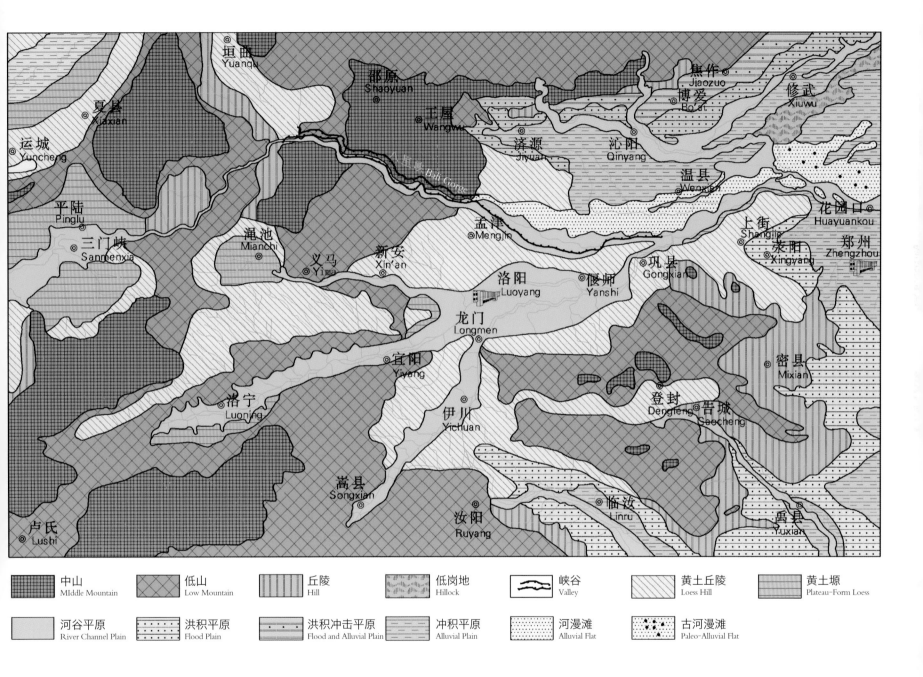

**黄河三门峡—八里峡段区域地貌分区图** (1985年)

Geomorphological Division Map of the Yellow River between Sanmenxia and Bali Gorge (1985)

大河日出  
River Sunrise

王屋春意  
Wangwu Spring

大美三峡  
The Beautiful Three Gorges

大峪湾  
Dayu Gulf

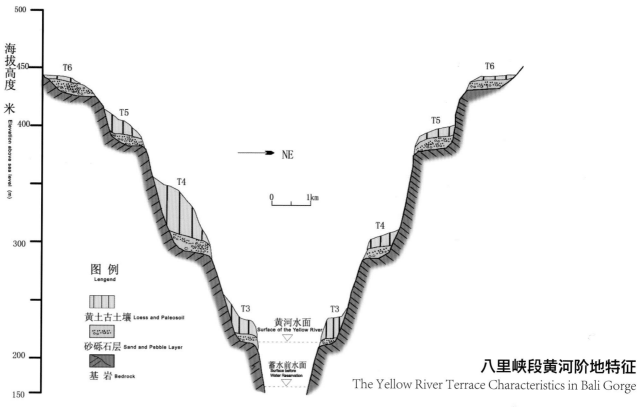

**八里峡段黄河阶地特征**
The Yellow River Terrace Characteristics in Bali Gorge

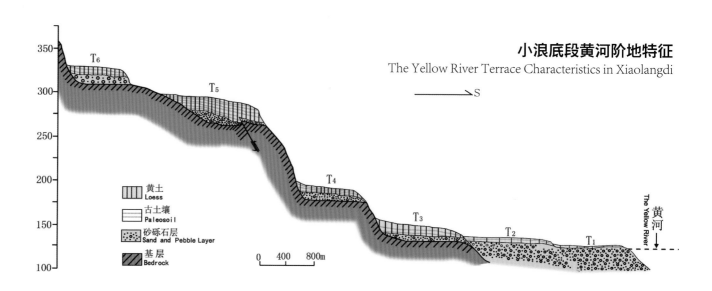

**小浪底段黄河阶地特征**
The Yellow River Terrace Characteristics in Xiaolangdi

# V 龙潭峡景区
## Longtan Valley Scenic Area

龙潭峡是一条由紫红色石英砂岩经流水追踪下切形成的V型峡谷，有"天然的波痕陈列馆"之称。峡谷全长12千米，谷内嶂谷、隘谷呈串珠状分布，云蒸霞蔚，激流飞溅，红壁绿荫，悬崖绝壁，不同时期的流水切割、旋蚀、磨痕十分清晰，巨型崩塌岩块形成的波痕大绝壁、"天碑石"国内外罕见。

Longtan Valley Scenic Area is a V-shaped valley formed by the red quartz sandstone after the fluvial incision, as is called the 'natural museum of ripple marks'. The overall length of the valley is 12km, with narrow valley and screen valley distributed like a string of bead. Colorful and flourishing, flying rip current, red cliff and green shade, steep precipices and cliffs coexist in the valleys. Fluvial incision, erosion, and polishing scratch in different periods are clearly, big ripple mark precipice formed by the huge collapsed blocks, Natural Stele Rock are rare in the world.

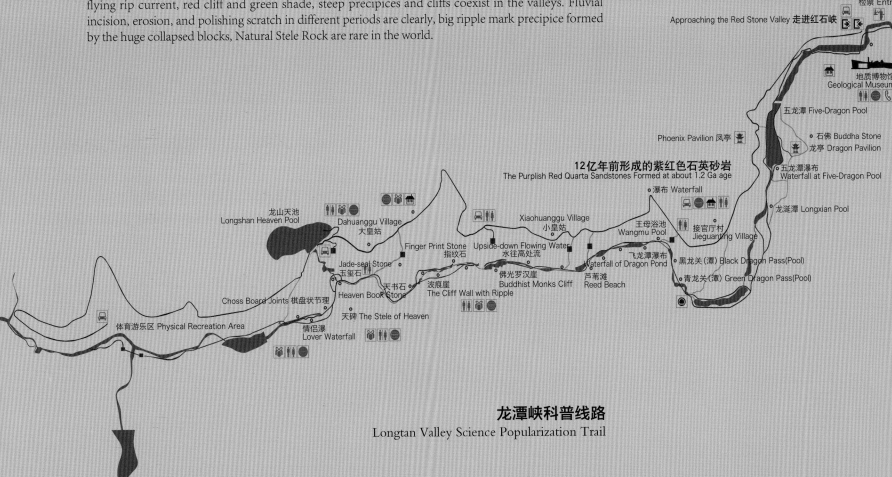

**龙潭峡科普线路**
Longtan Valley Science Popularization Trail

红岩嶂谷
Red-stone Cliffs and Narrow Valley

天碑
The Stele of Heaven

天碑位于龙潭峡的尽端，是一块超过50米高的巨型崩塌岩块。从不同角度仰望此石，或刀背、或苍鹰、或天碑。它一景多变，移步换景，构成景区一道极为靓丽的风景线。

The Stele of Heaven at the end of Longtan Valley is a peerless 'monument' created by nature. It has a height of more than 50m, formed by rock collapse. Seen from different point of view, the scape also changes. It presents itself in the shape of a knife blade, an eagle or a stele, all varying with your imagination.

小桥流水
Small Bridge and Trickling Water

天池
Heaven Pool

Scarp Formed by Revolving Erosion
旋蚀崖壁

Splendid and Charming Red-stone Cliffs and Narrow Valleys
多姿多彩的红岩嶂谷

水流潺潺
Murmuring Brook

俯视龙潭峡
Looking down into Longtan Valley

一线峡
The Thread Valley

**飞桥横卧**
Spanning Bridge across the Deep-cutting Valley

**龙潭春色**
Longtan Valley Spring Scene

波浪石坪
Corrugated Stone Ground

波浪崖
A Scarp Densely Covered with Ripple Marks

Heaven Book Stone 天书石

Arhat Buddha Cliff 佛光罗汉崖

Ripple Stone 波痕石

Septarium Stone 龟背石

# VI 黛眉山景区
## Daimeishan Scenic Area

黛眉山风光
Scenes in Daimeishan

黛眉山景区位于小浪底水库上游南岸的河南省洛阳市新安县北部，北临母亲河万里黄河与山西垣曲相望，西隔金陵涧水与荆紫仙山为界，南群岭诸峰远眺崤山，面积约47平方千米。流水沿12亿年前的紫红色石英砂岩中发育的两组近于直交且连通性好的垂直节理深度下切，形成了"两岸伟岩半空起，绝壁相对一线天"的红岩嶂谷、"山顶平缓如台，四周为断崖围限"的方山地貌和"如塔如城如台"的奇峰景观，由于崩塌而揭露的交错层理、波痕、泥裂等地质遗迹再现了古海洋潮涨潮落壮观场面，具有很高的研究价值和观赏价值；山顶数百亩的高山黛眉寨草甸为距今500万~260万年形成的夷平面，林木茂密，花卉丛生，水草丰美，共同构成一幅雍容华贵的丹青画卷。

Daimeishan Scenic Area is located at the bank of the upstream of Xiaolangdi reservoir, the north part of Xin'an County, Luoyang City, Henan Province. The mother river, the Yellow River is in the north and faces to Yuanqu, Shanxi Province. The west is Jinling stream, Which is the boundary between Jingzixianshan. Mountain ranges and peaks in the south look far into the distance with Xiaoshan, covering an area of 47km².

The flow deep cut along the vertical and well connected vertical joints developed by the red quartz sandstone 1.2Ga ago, and formed magnificent red stone valley, mesa landform and peak landscape. The geoheritages as cross bedding, ripple marks, mud cracks, which are revealed after collapse bring the spectacular scene of rise and fall of the tide in ancient ocean into life, with high value on research and appreciation; Hundreds of mu of Daimei meadow on the mountain top is the planation surface formed 5 to 2,6Ma ago from now, where there are thick forest, clustered flowers and abundant water, constituting an elegant and exalted picture.

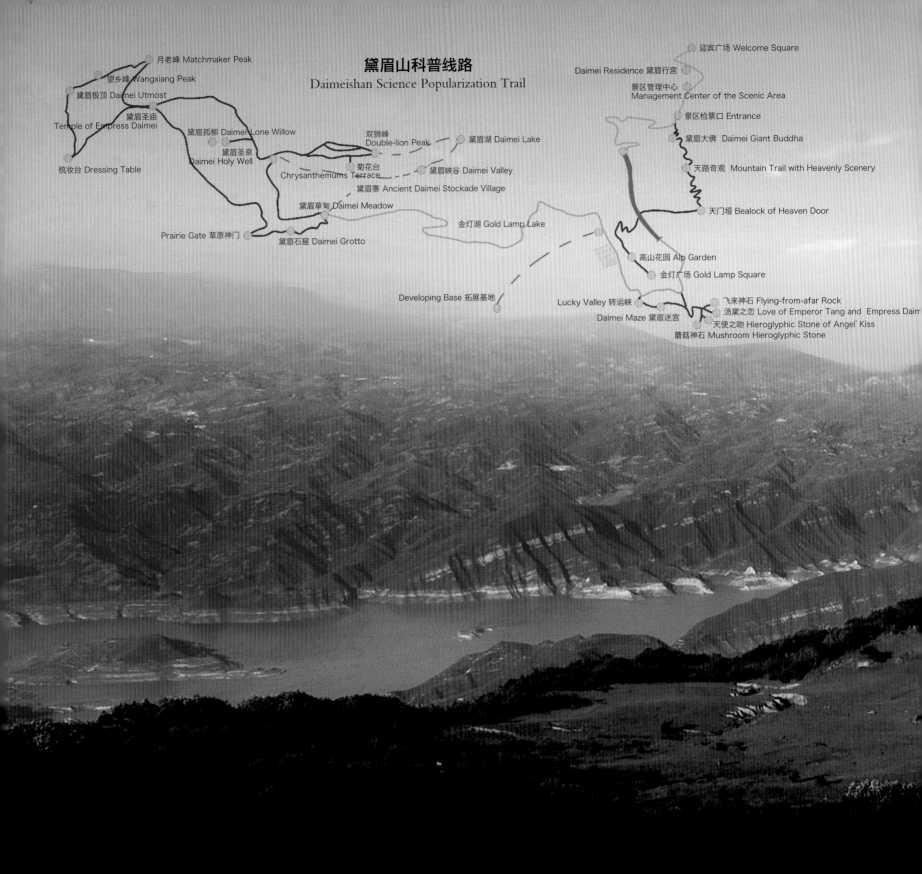

高山草甸
Alpine Meadow

黛眉山风光
Scenery of Daimeishan

天路奇观
Pilgrim's Wonders

黛眉春色
Daimeishan in Spring

地质画廊
Geological Gallery

草原神门
Prairie Gate

飞来神石
Flying-from-afar Rock

白雪皑皑的黛眉山犹如仙境，素裹的林木像洁白的珊瑚，修长的苇草暴露在雪外，阳光照过，像穿了金色的衣服。红色的山石在光照下泛着红润的光泽，那丰满的线条韵味十足，或许只能用"绕梁三日"来形容了。

  Silver woods Daimeishan looks like a fairyland, wrapped woods looklike white corals, slender reed exposed or stretched out of snow. Sunlight casts upon the slender reeds, as if putting on a golden clothes on them. Red rocks are blazing a rosy gloss and the beautiful texture of the rock will surely leave you with impression beyond words.

# VII 荆紫仙山景区
## Jingzixianshan Scenic Area

荆紫仙山景区位于黄河南岸，主要地质地貌景观为在 5 亿年前沉积的灰岩上形成的台阶状悬崖峭壁。景区主要景点有荆紫生岚、九龙戏水、石屏叠翠、荆山红叶、关郎春晴、阆苑牡丹、牧马古地、桃园春色等大八景，还有礼斗天坛、奠圣险崖、薰池穴居、山神望月、天门通衢、金台观日、水洞仙迹、登天云路等小八景。

Jingzixianshan Scenic Area is located at the south bank of the Yellow River, and the main geological landscape is step-shaped precipice formed on the limestone which deposited 500Ma ago. The scenic area has eight big scenic spots, namely, Jingzi foggy scene, Jiulong Xishui, Shiping Diecui, Jingzi red leaves, Guanlang Chunqing, Langyuan peony, Muma Gudi, Spring peach garden. In addition, there are also eight small scenic spots, namely, Lidou Temple of Heaven, Diansheng steep cliff, Xunchi cave building, Watching moon mountain god, Tianmen Tongqu, Viewing sun on Jintai, Immortal trace in the water cave, and Cloud road to heaven.

**荆紫仙山科普线路**
Jingzixianshan Science Popularization Trail

荆紫仙山雪景
Snowy Jingzixianshan Scene

漫山花海
Flowers all over the Mountains

荆紫仙山主峰
Main Peak of Jingzixianshan

雾锁荆紫
Mist-shrouded Jingzixianshan

荆紫仙山雪景
Snowy Jingzixianshan Scene

"荆紫生岚"为新安县古时八大景之一。荆紫仙山四周群山耸峙，北临黄河，壁立万仞，是观赏黄河小浪底两岸风光的绝佳景点。它是大黄河旅游带上一颗璀璨的明珠，是一处集道教文化、观光游览、生态休闲、度假疗养、户外运动、文化研修为一体的综合型景区。

'Jingzi Foggy Scene' is one of the ancient eight landscapes in Xin'an Country. Jingzixianshan is close to the Yellow River from north. Its steep cliff is surrounded by precipitous mountains. It is a great attraction to watch the scenery of both sides of Xiaolangdi Dam. It is a bright pearl along the Yellow River tourism belt. The scenery area integrates Taoism Culture experiencing, sightseeing, eco-leisure, holiday rest, outdoor activities and culture learning.

荆紫仙山云海
Cloud Sea on Jingzixianshan

日映云海
The Sunshine upon the Clouds

# Ⅷ 公园人文景观
## Cultural Resources in the Geopark

济渎庙石碑
Stone Tablet of Jidu Temple

公园地处黄河流域、中原地带，人文历史源远流长、光辉灿烂。远古的王屋山，是始祖文化的发源地，留下了避秦沟新石器遗址和女娲补天、黄帝祭天、愚公移山等动人的传说；中古的王屋山，成为道教圣地，被誉为"天下第一洞天"；现代的王屋山，以一代伟人毛泽东的"愚公移山、改造中国"而唱响全国，愚公精神成为中华民族之魂。在人类文明的历史文化长河中，王屋山处处闪烁着璀璨的光辉。

The geopark is located in the Yellow River basin, the center of China, with rich culture and long history. The ancient Wangwushan, is the headstream of Shizu culture, where we found the neolith site of Biqingou, and we heard the legends of Nuwa mending the sky, Emperor Huang sacrificing to heaven, Yugong moving mountains; In mediaeval period, Wangwushan became the holy land of Taoism, which is called 'the first cave paradise'; In modern times, it's famous in China for the word 'Reform China relentlessly with Yugong's spirit' from Chairman Mao, and Yugong spirit become our national soul. In the long history of Chinese splendid culture, Wangwushan will always shine brightly.

汉代函谷关遗址下发掘出的"关"字瓦当
Tiles of the Word "关" Shape of Han Dynasty Excavated beneath Hanguguan Ruins

"愚公移山"是中国一个古老的寓言故事，讲述的是中国一位老人，为了改变自己的居住环境，带领子孙挖山不止，意图搬走横亘在家门口的王屋、太行两座大山。这个故事表达了古代劳动人民改造自然的美好愿望。

Yugong (the ambitions old man) who tried to remove two mountains—Wangwushan and Taihangshan is a famous folk story of ancient China. It talks of an old man, in the aim of improving the living conditions of his family, leading his offspring to dig Wangwushan and Taihangshan lying in front of his house with perseverance. The story showed the ancient people had been of industrious and courageous to improve the residential conditions.

王屋山祭天仪式
Worship Ceremony in Wangwushan

王屋极顶天坛阁
Tiantan Pavilion at the Top of Wangwushan

不老泉
The Spring of Immortality

天坛山顶
Tiantanshan Top

盘古寺
Pangu Temple

迎恩宫
Ying'en Temple

济渎庙小北海
Xiaobeihai Pool of Jidu Temple

济渎庙古建筑
Ancient Construction of Jidu Temple

千唐志斋内珍藏的碑帖
Steles Preserved in Qiantangzhizhai Museum

济渎庙
Jidu Temple

阳台宫三檐三层琉璃玉皇阁,为河南省最高大的古阁,高近20米,五踩云龙斗拱参差层叠,云带缠绕,规模宏伟。
Yuhuang Pavilion with three eaves and three layers glass in Yangtai Palace is the highest ancient pavilion in Henan Province. It is nearly 20m high. Its colorful brackets with clouds and dragons uneven and cascade, constituting a magnificent scale.

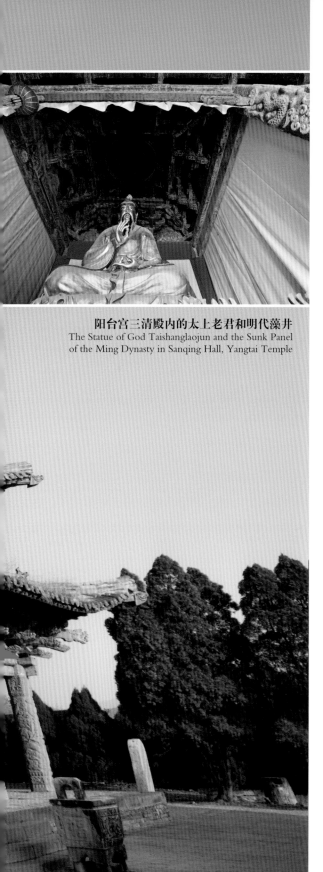

阳台宫三清殿内的太上老君和明代藻井
The Statue of God Taishanglaojun and the Sunk Panel of the Ming Dynasty in Sanqing Hall, Yangtai Temple

玉皇阁浮雕龙柱
Sculptured Dragon Post in Yuhuang Pavilion

新安汉函谷关遗址
Hanguguan Ruins of Han Dynasty in Xin'an Country

荆紫仙山通仙观
Tongxian Temple, Jingzixianshan

汉函谷关上联
The First Line of the Couplet of Hanguguan of the Han Dynasty

黄河小浪底大坝建于1997年11月。全长1667米，顶宽15米，底宽864米，是目前世界上规模最大的黏土斜心墙堆石坝。它总投资420亿元，创三项世界纪录，六项中国之最，在其上游形成了272平方千米的浩瀚水域，山水交融，水光潋滟。它不仅是伟大的水利工程，而且还造就了高峡平湖的壮丽景观。

　　Constructed in Nov. 1997, the Great Xiaolangdi Dam on the Yellow River is 1667m long, 15m wide on the top and 864m wide on the base. With a total investment of 42 billion RMB, the dam is the largest one of its types in the world. The construction of the dam sets up three world records and six domestic records. The hydrojunction is not only a hydraulic engineering but also the creates of the magnificent landscape of the tranquil lake among highly-elevated valleys. A vast water area of 272km² appears along the upper reaches.

**愚公渠东方红渡槽**
The Flume of the Yugong Irrigating Ditch

**小浪底大坝**
Xiaolangdi Reservoir Dam

# IX 公园生态资源
## Ecological Resources in the Geopark

公园丰腴的地力培育了丰富的生态资源，植物种类1000余种，被誉为南太行种子植物的基因库。王屋山原始森林和原始次生林面积达30平方千米。五角枫、红豆杉、银杏树、山白树、领春木等大量古近纪、新近纪孑遗植物成群分布、荟萃其间。春秋季节，百花盛开，红的、白的、紫的，如缎似锦把王屋山覆盖，绵绵无际。

Rich ecological resources are also developed on this fertile land. The Wangwushan region harbors 30km² of virgin forests and secondary forests, and 1,000 species of plants, and is therefore reckoned as the gene storehouse of seed plants in the southern Taihang Mountains. Various rare plant species are inherited from Paleogene and Neogene, such as Acer mono, Taxus chinensis, Ginkgo, Sinowilsonia henryi, and Euptelea pleiosperma sprinkle in the vegetation of this area. During spring days, hundreds of flowers bloom at the same time, covering the Wangwushan like limitless expansion of beautiful and bright silk.

生态绿野
Rich Vegetation

**千年银杏**
Ancient Apricot Tree

**阳台宫娑罗树**
Sal Tree in Yangtai Temple

# 黛眉古柏
Daimei Ancient Cypress

在龙潭峡谷口外，有一棵千年古檀，傍崖而立，高约30米，冠幅约40米，底部分四枝，集簇向上，胸围6.5米；树根裸露抓伏在岩石上，纵横交错，如盘龙卧虬，十分奇特。

There's a 1000-year-old ancient sandal tree stands towers at the mouth of Longtan Canyon. The tree growing beside the cliff is 30m tall with its crown breadth measuring up to 30m, has a width of about 40m tree crown. The base of the tree is divided into four branches,growing upper and then joining together growing upward. It is 6.5m in circumference and its roots expose to catch closely on the rock, criss crossing like coiled dragon and crouching tiger and is very strange.

在王屋山的深山密林里，还生活着18群3000多只太行猕猴，它是中国乃至世界分布最北的猕猴群，这些山中精灵，憨态可掬，嬉戏顽皮，给游客带来无限的情趣。

In Wangwushan, there inhabits 3,000 Taihang macaques in 18 groups which are the northernmost macaques. They are so cute and and attractive.

太行猕猴
Taihang Macaque

五龙口十里画廊
Ten-Mile Gallery in Wulongkou

# X 公园旅游
## Geopark Tourism

特色小吃——土炒馍
Special Snack——Fried Buns in Hot Soil

野酸枣汁
Wild Sour Jujube Juice

薄皮核桃
Thin Shelled Walnuts

王屋山灵芝
Glossy Ganoderma in Wangwushan

**冬凌产品**
Dongling Cool Tea Products

**新安县樱桃**
Cherries, Xin'an Country

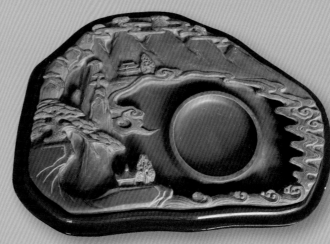

**王屋山天坛砚**
Tiantan Inkstone in Wangwushan

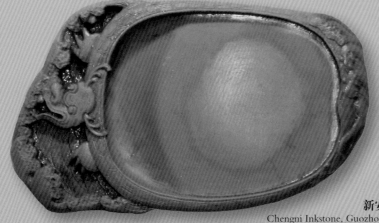

**新安县虢州澄泥砚**
Chengni Inkstone, Guozhou, Xin'an Country

王屋山—黛眉山地质博物馆
Wangwushan-Daimeishan Geological Museum

**黛眉山地质博物馆**
Daimeishan Geological Museum

**王屋山—黛眉山地质公园与西班牙维约尔卡斯—伊博尔—哈拉地质公园工作人员交流地学知识**
Exchanging Geoscience Knowledge between Wangwushan-Daimeishan Geopark and Villuercas Ibores Jara Geopark of Spain

**西班牙维约尔卡斯—伊博尔—哈拉地质公园考察小沟背景区**
Staff from Villuercas-Ibores-Jara Geopark of Spain Visiting Xiaogoubei Scenic Area

**西班牙维约尔卡斯—伊博尔—哈拉地质公园考察王屋山景区**
Staff from Villuercas-Ibores-Jara Geopark of Spain Visiting Wangwushan Scenic Area

游客如潮
Booming Tourism

西班牙维约尔卡斯—伊博尔—哈拉地质公园考察龙潭峡景区
Guests from Villuercas-Tbores-Jara Geopark of Spain Visiting Longtanxia Valley Scenic Area

篝火晚会
Campfire Party

普及科普知识
Popularization of Science Knowledge

地学科普活动
Geological Science Popularization Activity

地质科普夏令营活动
Geological Science Popularization Summer Camp

科普志愿者参观龙潭峡景区
Popular Science Volunteers Visiting Longtan Valley Scenic Area

小沟背女娲文化旅游节
Nuwa Cultural Tourism Festival in Xiaogoubei Valley

地球日小学生绘画
Primary School Students Painting on Earth Day

黄河三峡野玫瑰节
Wild Rose Festival in the Yellow River Three Gorges

王屋山登山节
Climbing Festival in Wangwushan

**编辑委员会**

**主　任**
李拴根　　济源市人民政府副市长
王琰君　　洛阳市人民政府党组成员、洛阳市政协副主席

**副主任**
冯正道　　济源市人大常委会副主任、党组成员
牛友谊　　济源市人民政府副秘书长

**成　员**
邱建平　　济源市国土资源局党组成员、副局长（主持工作）
冯长春　　济源市国土资源局党组成员、副局长
杨建统　　新安县国土资源局局长
刘晓玲　　王屋山—黛眉山世界地质公园管理局局长
陈万兵　　新安县黛眉山世界地质公园管理处主任
张忠慧　　河南省山水地质旅游资源开发有限公司副总经理
任利平　　河南省山水地质旅游资源开发有限公司高级工程师
罗自新　　河南省山水地质旅游资源开发有限公司高级工程师
陈新峰　　新安县黛眉山世界地质公园管理处办公室主任
李运生　　济源市国土资源局地质环境科科长

**策　划**　邱建平　冯长春　杨建统　刘晓玲　陈万兵
　　　　　张忠慧　任利平　罗自新　梁凯

**主　编**　景志刚
**执行主编**　程寰　刘鹏飞
**撰　稿**　刘晓玲　张忠慧　陈万兵　梁凯
**摄　影**　济源市摄影家协会
**装　帧**　周琰
**翻　译**　井燕　河南省山水地质旅游资源开发有限公司

**制　作**　《资源导刊》杂志社

## Editorial Committee

### Director
| | |
|---|---|
| Li Shuangen | Deputy mayor of Jiyuan government |
| Wang Yanjun | Member of the party group of Luoyang government, vice-president of Luoyang Political Consultative Conference |

### Deputy director
| | |
|---|---|
| Feng Zhengdao | Deputy head of the Standing Committee of the People's Congress of Jiyuan City, member of the party group |
| Niu Youyi | Deputy Secretary-General of Jiyuan Municipal Government |

### Member
| | |
|---|---|
| Qiu Jianping | Member of the party group and deputy director of Jiyuan Bureau of Land and Resources |
| Feng Changchun | Member of the party group and deputy director of Jiyuan Bureau of Land and Resources |
| Yang Jiantong | Director of Bureau of Land and Resources of Xin'an county |
| Liu Xiaoling | Director of the Administration of Wangwushan-Daimeishan UNESCO Global Geopark |
| Chen Wanbing | Director of the management office of Xin'an Daimeishan Global Geopark |
| Zhang Zhonghui | Deputy director of Henan Shanshui Geological Tourism Resources Development Co,. Ltd. |
| Ren Liping | Senior engineer of Henan Shanshui Geological Tourism Resources Development Co,. Ltd. |
| Luo Zixin | Senior engineer of Henan Shanshui Geological Tourism Resources Development Co,. Ltd. |
| Chen Xinfeng | Office director of the management office of Xin'an Daimeishan Global Geopark |
| Li Yunsheng | Chief of Environmental Department of Jiyuan Land Resource Bureau |

**Planning**: Qiu Jianping, Feng Changchun, Yang Jiantong, Liu Xiaoling, Chen Wanbing, Zhang Zhonghui, Ren Liping, Luo Zixin, Liang Kai

**Editor-in-chief**: Jing Zhigang
**Managing editor**: Cheng Huan, Liu Pengfei
**Contribution**: Liu Xiaoling, Zhang Zhonghui, Chen Wanbing, Liang Kai
**Photography**: Jiyuan Photographers Society
**Designer**: Zhou Yan
**Translator**: Jing Yan  Henan Shanshui Geological Tourism Resources Development Co., Ltd.

**Producer**: Henan Land & Resources Herald Magazine Agency

# 图书在版编目(CIP)数据

锦绣山河:中国王屋山—黛眉山世界地质公园/王屋山—黛眉山世界地质公园管理委员会编. —郑州:黄河水利出版社,2016.5
 ISBN 978-7-5509-1432-2

Ⅰ.①锦… Ⅱ.①王… Ⅲ.①地质-国家公园-介绍-河南省 Ⅳ.①S759.93

中国版本图书馆CIP数据核字(2016)第102929号

---

组稿编辑:王路平  电话:0371-66022212  E-mail:hhslwlp@126.com

| | |
|---|---|
| 出 版 社: | 黄河水利出版社 |
| | 地址:河南省郑州市顺河路黄委会综合楼14层  邮政编码:450003 |
| 发行单位: | 黄河水利出版社 |
| | 发行部电话:0371-66026940、66020550、66028024、66022620(传真) |
| | E-mail:hhslcbs@126.com |
| 承印单位: | 河南省瑞光印务股份有限公司 |
| 开   本: | 787 mm×1 092 mm  1/12 |
| 印   张: | 10.5 |
| 字   数: | 250千字          印数:1—1 000 |
| 版   次: | 2016年5月第1版   印次:2016年5月第1次印刷 |

定价:128.00元